Andreas Eichler · Lucía Zapata-Cardona

# Empirical Research in Statistics Education

Andreas Eichler
Institute for Mathematics
University of Kassel
Kassel
Germany

Lucía Zapata-Cardona
College of Education
University of Antioquia
Medellín, Antioquia
Colombia

DOI 10.1007/978-3-319-38968-4

# Main Topics You Can Find in This ICME-13 Topical Survey

In this ICME-13 Topical Survey, we provide a review of recent research into statistics education. We focus our review on the empirical research that has been published in established educational journals or the proceedings of important conferences that include at least a section referring to statistics education. We have identified and will address six important research topics, namely teacher knowledge, teachers' statistics-related affect, teacher preparation, student knowledge, students' statistics-related affect, and student learning of statistics with technology. For each research topic we build upon existing reviews and add more recent research. In each section we start with a review of recent research and end with a brief conclusion for each research topic.

# Contents

# Empirical Research in Statistics Education

In this ICME-13 Topical Survey, we provide a review of recent research into statistics education. We focus our review on empirical research that has been published in established educational journals or the proceedings of important conferences that include at least a section referring to statistics education. We have identified and will address six important research topics, namely, teacher knowledge, teachers' statistics-related affect, teacher preparation, student knowledge, students' statistics-related affect, and student learning of statistics with technology. For each research topic we build upon existing reviews and add more recent research. In each section we start with a review of recent research and end with a brief conclusion for each research topic.

## 1 Introduction: Setting the Field

Statistics is a hot topic in modern society. Varian (2009), who was interviewed as the chief economist of Google in 2009, described the importance of statistics expressively:

> I keep saying the sexy job in the next 10 years will be statisticians. People think I'm joking, but who would've guessed that computer engineers would've been the sexy job of the 1990s? The ability to take data—to be able to understand it, to process it, to extract value from it, to visualize it, to communicate it—that's going to be a hugely important skill in the next decades, not only at the professional level but even at the educational level for elementary school kids, for high school kids, for college kids. Because now we really do have essentially free and ubiquitous data. So the complimentary scarce factor is the ability to understand that data and extract value from it.

In fact, polls have an impact on the distribution of financial resources. Statistical analysis of tests is the basis of medical progress. Statistical data govern decision making in economics, politics, and society. Due to the tremendous importance of statistics for numerous parts of our life, statistics including data analysis and



A. Eichler and L. Zapata-Cardona, *Empirical Research in Statistics Education*, ICME-13 Topical Surveys, DOI 10.1007/978-3-319-38968-4_1

probability have become a crucial topic in mathematics education in recent decades (e.g., Batanero et al. 2011). Conversely, education without statistics has become inconceivable, since "citizens who cannot properly interpret quantitative data are, in this day and age, functionally illiterate" (Mathematical Science Education Board and National Research Council 1990, p. 8).

Accordingly, in the past two decades, statistics learning and teaching has become a field of increasing educational research. Shaughnessy (2007, p. 957) already noted some years ago the "amazing boom in research, curriculum development, and assessment in statistics education" that makes it unfeasible to review the entire body of research in this field. For this reason, we restrict our focus considerably when addressing the current state of research in statistics education. First, we restrict our focus to empirical research. Second, we do not try to draw an exhaustive picture of the empirical research in statistics education since there have been excellent reviews in recent years. For example, Shaughnessy's (2007) review focused especially on the research in statistics learning before 2007. Moreover, Batanero et al. (2011) edited volume which presents the worldwide status quo of research on statistics teachers' knowledge and beliefs from the ICME/IASE study. For this reason, we build upon existing reviews aiming to add more recent trends in research in statistics teaching and learning.

The main frame that we use for integrating recent empirical research in statistics education is the work of Gal (2002) concerning the construct of statistical literacy. We first use the distinction between an individual's knowledge and disposition when we consider research on teacher knowledge and teachers' statistics-related affect (Hannula 2012). We add to these two research topics a review of research into teacher preparation and professional development. Afterwards we review recent research on learners of statistics. Again, we use the distinction between knowledge and disposition when we consider research on student knowledge and research on students' statistics-related affect. Since technology is an important aspect of research in statistics education, we add a specific section addressing research on learning statistics with technology. Before we target different aspects of research, we briefly outline the method of gaining relevant research papers and focus on the distinction between knowledge-related and dispositional elements of statistical literacy.

## 2 Survey of the State of the Art

### 2.1 Sources for Research in Statistics Education

To identify reports focusing on the mentioned six research topics for this survey, we searched different educational journals and conference proceedings. More specifically, we searched relevant research reports in:

- the proceedings of the International Conference on Teaching Statistics (ICOTS),
- the proceedings of the Conference of the European Society for Research in Mathematics Education (CERME),

- the proceedings of the International Conference Turning Data into Knowledge,
- the *Statistics Education Research Journal*,
- the *Journal of Statistics Education*,
- *Educational Studies in Mathematics*,
- *ZDM Mathematics Education*, and
- the *Journal of Mathematics Teacher Education*.

In these publications we searched relevant literature using keywords. For example, we used the keywords "teachers," "attitudes," "beliefs," and "statistics" to find relevant literature referring to statistics teachers' affect.

Further, we mostly restricted this search to existing reviews. Thus, we based our search concerning teachers on the ICME/IASE studies published by Batanero et al. (2011) and Shaughnessy (2007). We based the search about students' knowledge on Shaughnessy (2007) and the search about students' knowledge and affect referring to technology on Biehler et al. (2013). Finally, we searched more generally about students' statistics-related affect, because this aspect is not discussed in detail in the abovementioned reviews.

## *2.2 Knowledge and Dispositional Aspects of Statistical Literacy*

There seems to be an implicit consensus about the distinction between knowledge and disposition. By contrast, the literature provides many different meanings and descriptions of constructs such as affect, attitudes, beliefs, or motivation. Even concerning one of these constructs, beliefs, Fives and Buehl (2012, p. 471) stated that "research on teachers' beliefs … runs the gamut of research methodologies, theoretical perspectives, and identification of specific beliefs about any number of topics." For this reason, we refer to Hannula's (2012) description of an individual's (mathematics-) or statistics-related affect to distinguish first of all between knowledge and dispositional aspects that according to Gal (2002) represent an individual's statistical literacy. However, we refer in particular to Hannula (2012) to distinguish between different constructs within the dispositional aspect. Following Hannula (2012, p. 144), who describes an individual's affect as constituted by three "explanatory factors of behavior and learning," i.e., cognition, motivation, and emotion, we briefly address these three constructs in the next paragraphs.

### 2.2.1 Cognition

Following Philipp (2007), cognition includes knowledge and beliefs or rather propositions that have a truth value. Although the truth value could be used to distinguish between knowledge and beliefs, the distinction of these two cognitive aspects is difficult (Philipp 2007), since both aspects seem to be "inextricably intertwined" (Pajares 1992, p. 325). For example, a definition of knowledge as a proposition that is true independently from individuals is not possible if the epistemological model of constructivism (von Glasersfeld 1993) is applied. However, for this paper, we intend to make a pragmatic distinction between knowledge and beliefs: First, we regard individuals' propositions referring to statistical (or mathematical) concepts as students' or teachers' knowledge. Second, we use a helpful strategy of Philipp (2007) to distinguish between knowledge and beliefs: An individual could accept a disagreement to a belief, but not to knowledge. For example, "probability is a function with specific characteristics" could be understood as an individual's knowledge, while it is possible to accept that a belief such as the proposition "statistics is a tool to solve real world problems" could be more or less true for different persons. Thus we use Philipp's (2007, p. 259) definition of beliefs as "psychologically held … propositions about the world that are thought to be true".

### 2.2.2 Motivation

Following Hannula (2006, p. 167), motivation consists of parts of cognition and parts of emotion since "for example, the motivation to solve a mathematics task might be manifested in beliefs about the importance of the task (cognition), but also … in sadness or anger if failing (emotion)." Similarly, Rheinberg et al. (2001) use both cognitive and emotional aspects to define and measure a current motivation, including, for example, the aspect of interest. Beliefs about self that are investigated as self-concept or self-efficacy could especially be understood as part of motivation (Watt and Richardson 2015). Values or goals could also be understood as being both part of motivation (Watt and Richardson 2015) and a specific form of an individual's system of beliefs (Philipp 2007; Eichler 2011). We later address teachers' goals when discussing statistics teachers' motivation and refer to self-efficacy and values when discussing statistics teachers' beliefs.

### 2.2.3 Emotion

Emotional dispositions could be understood as attitudes (Hannula 2012). Thus, to differentiate beliefs and attitudes, it is possible to define attitudes as "emotional dispositions towards mathematics" and "perceived competence in mathematics" (Di Martino and Zan 2010, p. 44) that are not based on propositions that are true or false. For example, the attitude "I like statistics" has no logical value, but shows "a psychological tendency that is expressed by evaluating a particular entity with some degree of favor or disfavor" (Eagly and Chaiken 1998, p. 270). In contrast, the belief "statistics is a tool to solve real world problems" has a logical value and could be assigned a value of true or false. The theoretical distinction between different aspects of learners' and teachers' knowledge elements and dispositional elements as parts of their statistical literacy is used in the following sections to discuss existing research in statistics education.

## 2.3 *Teachers' Knowledge of Statistics*

Shaughnessy's (2007) literature review addressed also teachers' statistical knowledge. One of his conclusions was that there should be a shift in the way scholars look at teachers' knowledge, where the focus should be on knowledge that is context related and relevant to the teachers' daily practice rather than exclusively on content knowledge, which has for a long time been the center of research on teachers' knowledge. The context relatedness of research on teachers' knowledge mostly frames this review section.

### 2.3.1 Relation of Statistics and Mathematics

Although much is known about teacher knowledge in relation to teaching mathematics, the situation for statistics is not as clear. What is statistical teacher knowledge? What is the knowledge teachers need to teach statistics? In spite of the fact that the mathematical knowledge needed for teaching and the statistical knowledge needed for teaching share some similarities, there are also some differences that respond to the uncertain, inductive, and subjective nature of statistics compared to mathematics (Cobb and Moore 1997). This debate has been informed by Groth (2007, 2013), who sketched a hypothetical descriptive framework of statistical knowledge for teaching inspired by the literature on mathematical knowledge for teaching and the *Guidelines for assessment and instruction in statistics education* (*GAISE*) *report* (Franklin et al. 2007). Burgess (2011), on the contrary, argued that the study of statistics teachers' knowledge cannot be carried out using the literature related to mathematics teachers' knowledge. Based on the statistical thinking in empirical research (Wild and Pfannkuch 1999), Burgess developed a theoretical framework to explore teachers' knowledge used in teaching statistics through investigations.

### 2.3.2 Statistical Knowledge of Prospective Teachers

The majority of the research studies have used frequent testing to assess participants who were prospective teachers enrolled in statistics courses or education courses at college level. Several studies illustrate this type of design. For example, Hannigan et al. (2013) studied the conceptual knowledge of 134 prospective secondary teachers whose statistical knowledge was assessed using the Comprehensive Assessment of Outcomes in Statistics (CAOS) test. They found that the prospective teachers performed poorly on items that included randomization, sampling and populations, and extrapolation from a regression model. Leavy (2006) studied the evolution of prospective primary teachers' knowledge about distributions in a graduate statistics course. The results showed that the prospective teachers at the beginning of the course relied exclusively on descriptive statistics measurements, and during the course they started to include graphical representations that gave details about the distributions as a complement to the descriptive statistics measurements. Casey and Wasserman (2015) explored teachers' knowledge of informal lines of best fit using task-based interviews. The participants were 11 pre-service and 8 in-service teachers enrolled in teacher education courses. The researchers found some significant gaps in their content knowledge. Further studies focusing on the statistical knowledge of prospective teachers refer to teachers' statistical literacy in general (Koleza and Kontogianni 2013) or refer to specific issues such as teachers' interpretation of central tendency measures (Santos 2013).

### 2.3.3 Statistical Knowledge of In-Service Teachers

Fewer studies have explored in-service teachers' knowledge. Research on in-service teachers' knowledge about statistics has used task-based interviews, tests, and observations of statistics teaching practice.

Jacobbe and Horton (2010) investigated three strong mathematics teachers' comprehension of data displays. The researchers observed, interviewed, and assessed the teachers to get insight into what they understood. They found that the teachers were proficient in answering straightforward questions related to data displays but unsuccessful with questions that assessed a higher level of graphical comprehension. Jacobbe (2012) studied the understanding of three in-service elementary school teachers with respect to the concepts of mean and median. Interviews with these teachers revealed that they did not have a good conceptual knowledge of those basic statistical concepts. Hobden (2014) studied the level of statistical literacy of 316 in-service non-mathematics teachers that would potentially teach mathematics in the near future. The participants were enrolled in a government-funded teacher development program, and the data were collected from the teachers' explanations of the concept of median in the context of HIV/AIDS survival times. The results revealed that the teachers had a poor understanding of the median. Kataoka et al. (2014) studied in-service teachers' understanding of covariation within a professional development program using a task that related height and arm span. They found that the teachers improved their understanding of covariation. Referring to the same professional development program, da Silva et al. (2014) studied teachers' understanding of variation using graphical representations such as dot plots and box plots. They found that the teachers knew how to compute variation measures but did not know how to analyze the values. Bansilal (2014) studied 290 in-service teachers' knowledge about normal distributions. The participants were enrolled in a teacher development program and were assessed with a task. Teachers' responses were analyzed using the Action, Process, Object, Schema (APOS) framework, and the results revealed that the teachers experienced problems linking the probability values with the area covered by the curve.

Particularly interesting are the studies from Casey (2010) and Peters (2011), who gathered information from in-service teachers to develop theoretical frameworks. Casey (2010) observed three experienced in-service teachers while they were teaching about the association between variables. With the information from about 50 class observations, Casey constructed a theoretical framework to describe the statistical knowledge for teaching. The findings revealed that teachers need a substantial knowledge base for teaching the concept of a correlation coefficient (computation, interpretation, sensitivity, estimation, and terminology). In contrast, Peters (2011) interviewed 16 experienced secondary statistics teachers while solving three variation tasks. With the data collected, Peters established a theoretical framework for developing robust understandings of variation.

### 2.3.4 Teachers' Context Related Knowledge

Arnold (2008) took a different perspective. Her study focused on in-service teachers' statistical knowledge but within a professional development program in learning communities. The teachers described areas of statistics in which they needed support, and the activities proposed were intended to fulfill those needs (sources of variation, gathering and cleaning data, comparison of sample distributions, and determining appropriate variables and measures). The focus of this research was not only on the teachers' conceptual knowledge but on improving their statistics teaching practice. The results revealed that, in spite of the improvement in teachers' knowledge, they still required ongoing support from the learning communities and within their practice. A further research approach was provided by Bakogianni (2015), who followed 10 secondary teachers by collaboratively developing and implementing a task for eighth graders as an introduction to statistics. She concluded that through collaboration the teachers' gained an increased understanding of statistical concepts and an increased awareness of students' difficulties.

### 2.3.5 Conclusion

There have been very few studies of teachers' content knowledge in relation to their teaching practice. Most studied teachers' knowledge in order to make strong cases for teachers' lack of disciplinary knowledge. Some of them focused mainly on how well teachers understood a statistics topic. However, since teachers' professional knowledge is a combination of multiple dimensions, it is simplistic to focus exclusively on teachers' content knowledge. Although strong statistical knowledge is required to teach statistics, the knowledge by itself is not enough. Thus, it is also important to know how the teacher uses that knowledge in teaching. As mentioned by Ponte (2011), teachers' knowledge is not exclusively declarative knowledge but action-oriented professional craft knowledge, which is essentially practical. The study by Hannigan et al. (2013) showed that the prospective teachers who had just completed a module in introductory statistics performed much better than those who had taken the module 1–2 years before the study, indicating that students forget what they do not use and that the instruction is only for the moment. This might be an indication that research should focus on what teachers do in their teaching with their statistical knowledge and how they exploit their statistical knowledge to serve their teaching.

The main reflections from this literature review are related to theoretical and methodological aspects of research on teachers' knowledge. From the theoretical point of view, it is clear that teachers' knowledge of statistics is primarily defined in terms of what teachers know about statistics. The community has to ask these questions: What is teachers' knowledge? Is what teachers know about statistics enough to define their knowledge? What is the knowledge teachers need to teach statistics? From the methodological point of view, an interesting amount of research has studied prospective teachers who either never have been in a classroom as student teachers or have not had full control of a class. For this reason, one of the main questions is whether research with prospective teachers really gives us information about teachers' knowledge.

## *2.4 Teachers' Statistics-Related Affect*

As described above, when looking at the dispositional part of teachers' statistical literacy, we distinguish between three aspects of teachers' statistics-related affect, i.e., teachers' attitudes, teachers' motivation, and teachers' beliefs. We discuss recent research on these three aspects based on reviews of Chick and Pierce (2011), Eichler (2011), and Estrada et al. (2011).

### 2.4.1 Instruments to Measure Teachers' Statistics-Related Affect

The most established instruments for measuring teachers' statistics-related affect primarily seem to aim at measuring statistics teachers' attitudes: the Statistics Attitudes Scale (SAS) by Roberts and Bilderback (1980), the Attitudes Towards Statistics (ATS) by Wise (1985), and the Survey of Attitudes Toward Statistics (SATS) by Schau et al. (1995). However, using the definition of the construct of attitudes of Eagly and Chaiken (1998, see above), these three instruments measure not only attitudes but also beliefs and motivation. For example, the item "I will enjoy taking statistics courses" (SATS, Schau et al. 1995) expresses a preference towards statistics, and thus an attitude towards statistics. By contrast, the item "statistics involves massive computations" (SATS, Schau et al. 1995) is a proposition that is individually false or true and is thus a belief. Finally, the statement "I am interested in using statistics" (SATS, Schau et al. 1995) represents a persons' interest as an aspect of motivation. Although the three instruments aim at measuring a different number of constructs, the items in all instruments refer to (a) attitudes (feelings about statistics), (b) beliefs (e.g., self-efficacy, according to Bandura 2012), and (c) motivation (e.g., interest; c.f. Rheinberg et al. 2001). For this reason, we use results that are gained through the mentioned three instruments to discuss research on emotions, motivation, and beliefs towards statistics.

### 2.4.2 Teachers' Attitudes Towards Statistics

Several researchers used the SATS to measure statistics teachers' attitudes—most of the investigated teachers were prospective teachers—as learners of statistics (Batanero et al. 2005; Chick and Pierce 2011; Hannigan et al. 2013; Nasser 2004). As a common result, they reported slight positive attitudes (feelings) concerning statistics. Using other instruments, Begg and Edwards (1999) also reported teachers' positive attitudes towards statistics. Interestingly, Onwuegbuzie (1998) found that the attitudes of prospective teachers about statistics are lower than the attitudes of other students. A similar result is found by Sturm (2016), who studied 64 prospective teachers. Hannigan et al. (2013) further found in a study of 134 prospective teachers that postgraduate students' attitudes were more positive than those of undergraduate students. This implicit effect of the maturation of students was also reported by Estrada et al. (2011).

Whereas researchers mostly agree about the status quo of statistics teachers' attitudes, research yielded differing results when referring to the effect of attitudes on achievement in statistics. For example, Hannigan et al. (2013) found no significant relation between attitudes and achievement. In contrast, Nasser (2004), in a study with 162 teachers, found a moderate positive correlation and, finally, Zientek et al. (2010), in a study with 95 participants, reported a strong impact of prospective teachers' attitudes on their achievement. A possible reason for the contradictory results is the different definition of achievement in the three studies. Further, an impact of the maturation on students' achievement could influence the different results (Hannigan et al. 2013).

Since the SATS measures not only attitudes but also beliefs and motivation, these studies mostly measured correlations between attitudes and more cognitive variables. Estrada et al. (2011, p. 167) stated that "liking or disliking statistics was related in these teachers to their perception of self-capacity to learn statistics and to the value given to statistics." Zientek et al. (2010) found these relations to be strong.

There are few studies that researched statistics teachers' attitudes in a qualitative design. Martins et al. (2012) analyzed the written responses of 175 in-service teachers to several items of a questionnaire that yielded low scores in a previous study. They found a variety of reasons for rating attitudes items positively or negatively. For example, the lack of motivation or a perceived lack of knowledge yielded a negative attitude towards teaching statistics. Thus, Martins et al. (2012) found motivation, knowledge, or beliefs as a reason for teachers' attitudes. Similarly, Leavy et al. (2013) reported several of the 134 teachers' rationales for holding positive or negative attitudes about statistics that concerned the nature of statistics or the role of the context.

In addition there are two research approaches that amongst others focus on the development of statistics teachers' attitudes. For this aspect, the results of Hannigan et al. (2013) and partly the results of Batanero et al. (2005) gave evidence about the development of statistics-related attitudes through maturation.

### 2.4.3 Teachers' Motivation and Statistics

A crucial aspect for motivation seems to be an individual's interest in a specific topic (Rheinberg et al. 2001; Watt and Richardson 2015). Although this aspect was investigated in studies using the SATS, it is not easy to interpret the results referring to the teachers' interest. In some studies (e.g., Hannigan et al. 2013) the teachers' ratings of interest towards statistics seemed to be lower than other variables. However, existing studies including the variable of interest did not focus on a comparison of teachers' interest to other variables of teachers' affect.

A further main aspect of researching motivational variables of teachers is to analyze teachers' goals (Watt and Richardson 2015). Using a qualitative study with 13 teachers and a quantitative study with 113 teachers, Eichler (2011) discussed two different overarching goals of statistics teaching, namely emphasizing statistics as applied mathematics developed in a process or statistics as static part of mathematics that is not necessarily related to an application. The research of Watson (2001)—including 43 in-service teachers and aiming to analyze profiles of teachers—suggested that the aspect of application of statistics in everyday life is the main reason for teaching statistics. A further approach to investigate statistics teachers' goals as a motivational variable is to analyze statistics teachers' learning orientation (c.f. Staub and Stern 2002). For this purpose, Zieffler et al. (2012) provided a questionnaire called the Statistics Teaching Inventory (STI) that is related to the aims of the GAISE report.

Apart from the research of Sturm (2016), who described a moderate change in teachers' interest in statistics based on a short-term intervention, we did not find research that focused on the development of teachers' motivation.

### 2.4.4 Teachers' Beliefs Towards Statistics

A main interest in researching teachers' beliefs is to investigate values towards statistics that Philipp (2007) describes as deeply held beliefs. For example, Begg and Edwards (1999) investigated the beliefs about the benefit of statistics for society of 34 in-service and prospective teachers, amongst others. These teachers emphasized statistics as a tool to understand real life or rather decision making in real life. Using the SATS to assess the value of statistics for society and personal life, Hannigan et al. (2013) stated that prospective teachers "placed a value on statistics," which is in agreement with other studies in this field. Sturm (2016) distinguishes further the value of statistics for society and the value of statistics in personal life. She found that prospective teachers valued the benefit of statistics for society high and even higher than the benefit for their personal life.

A further question in research on statistics teachers' beliefs is the nature of statistics as a discipline in or outside the domain of mathematics (Cobb and Moore 1997). Whereas several researchers claimed that statistics is different from mathematics, empirical studies mostly showed that teachers understood statistics as a part of mathematics (e.g., Begg and Edwards 1999; Chick and Pierce 2011). Further evidence for this belief was given by the strong correlation of scales measuring the teachers' interest in mathematics and statistics (Sturm 2016) and the strong correlation of scales measuring anxiety towards mathematics and statistics (Nasser 2004). A different qualitative study that included 50 teachers showed that teachers understood statistics as an integral part of mathematics and understood statistics as applied mathematics (Eichler and Erens 2015).

As reported above, teachers' beliefs about themselves measured by the SATS mostly gave evidence that the prospective statistics teachers were "confident about their intellectual knowledge and skills when applied to statistics" (Hannigan et al. 2013, p. 443). However, research that investigated the relation between statistics teachers' beliefs and these teachers' knowledge gave evidence that the teachers overestimate their statistical competence (Nasser 2004).

Compared to the review of Eichler (2011), research yielded no further results referring to the impact of statistics teachers' beliefs on classroom practice or student learning. Recapitulating this review, statistics teachers' beliefs seem to impact on these teachers' classroom practices, particularly if there is a differentiation between central and peripheral beliefs. Further, the relation between statistics teachers' beliefs and their students' knowledge and beliefs is vague, although there are results that imply an impact of the teachers' learning orientation related to a constructivist orientation or related to emphasizing real data on students' knowledge and beliefs.

Pearson's (2014) study was one of very few that addressed belief changes. In this research, the impact of a professional development program on 14 teachers was discussed. Results showed that it seemed to be possible in particular to increase the teachers' beliefs concerning the value of statistics. A similar approach by Sturm (2016) in a short-term intervention, however, yielded less evidence for a belief change. An interesting approach was reported by Olfos et al. (2014), who presented the belief changes of 28 teachers that were based on the ongoing discussion of lesson studies.

### 2.4.5 Conclusion

Our first conclusion is that when comparing research into teachers' knowledge and teachers' statistics-related affect the knowledge aspect of teachers' statistical literacy is much more addressed in statistics education research than is the dispositional aspect. However, existing research shows that teachers seem:

- to hold positive attitudes towards statistics,
- to attach a considerable value to statistics, and
- to perceive statistics particularly as a field of applied mathematics.

It is also striking that when comparing research into teachers' knowledge and teachers' statistics-related affect we found that in both fields of research there is a strong tendency to investigate prospective teachers. For this reason, there is a lack of research concerning the dispositional aspect of in-service teachers' statistical literacy in terms of attitudes, motivation, or beliefs. However, since there seems to be a consensus in mathematics education research that dispositional aspects such as beliefs are strongly context related (e.g., Skott 2009), there is a need to investigate teachers in their real professional context, i.e., the teachers' classrooms.

## 2.5 *Teacher Preparation in Statistics*

There is no doubt that teacher preparation is a factor that contributes to the quality of teaching of statistics. However, the research has not attended to statistics teacher preparation to the same degree as other topics in the field of teaching and learning statistics. Shaughnessy (2007) proposed that future research focus mainly on the knowledge teachers need to teach statistics. He recommended exploring questions such as "What is the statistical knowledge necessary for teaching?" Although answering these questions would be an interesting contribution to research on teacher knowledge, answering them would fall short in considering the overall tensions and constraints in preparing teachers to meet the challenges in teaching statistics. In this section, we discuss several research results that refer to prerequisites for teacher preparation in a broader sense. Different formats for research on statistics teacher preparation programs have been proposed in the literature. Some research has been carried out in college settings in which the participants are pre-service teachers; other research has been undertaken in the statistics classroom where in-service teachers conduct their practice. Some studies have focused on teacher content knowledge and others on strategies for helping teachers develop pedagogical skills to deal with the tensions in their classrooms. After outlining three paradigms that are implicitly or explicitly the basis for the abovementioned research formats, we discuss related research.

### 2.5.1 Paradigms of Teacher Preparation

There is an ongoing discussion about the different paradigms for teacher preparation. From a sociological point of view, the pedagogical models for preparing teachers can be situated within three paradigms: rationalist, naturalistic, and critical (Loya-Chávez 2008). A rationalist paradigm focuses attention on the centralization and normalization of knowledge. In this paradigm, the teacher becomes an intellectual who dominates scientific disciplines in order to teach them, and the teacher preparation programs follow the principles of technical rationality—the amount of professional knowledge determines a teacher's successful teaching (c.f. Schön 1992). In the naturalistic paradigm, the preparation for teaching is related to the understanding of the natural development of the learners, and this paradigm follows the line of research of Piaget, Montessori, and Decroly. The naturalistic paradigm is based on a deep understanding of the principles of learning, and teaching is centered on facilitating this learning. The critical paradigm suggests that teachers' preparation needs to reflect on their teaching practice (reflection in action). This paradigm promotes conditions where teachers critically deal with their educational reality in order to improve it. The teacher is an autonomous professional that reflects on the daily practice of teaching.

### 2.5.2 Research on Teacher Preparation in College Settings

Studies in college settings have centered their research interest on teacher preparation programs carried out within formal education at the tertiary level, primarily pre-service teacher programs, and have focused on how, within a statistics course required for a specific education degree, pre-service teachers develop certain skills, knowledge, or affect toward statistics. Most of the research in college settings follows a naturalistic paradigm of teacher preparation in which the researchers develop ways—supported by psychology—to facilitate the statistics learning of prospective teachers. For example, Dolor and Noll (2015), in the setting of an undergraduate statistical course, studied the learning trajectory of pre-service teachers when carrying out hypothesis tests. The researchers reported that the participants were able to generate an informal version of a hypothesis test for categorical data. Magalhães and Magalhães (2014), in the context of an introductory undergraduate statistics course, developed four activities designed to help pre-service teachers improve their initiative, collaboration, intentional action, and attitudes related to teaching statistics. Research with pre-service teachers is important because it informs the academic community about relevant results in terms of teachers' knowledge and learning. However, research with pre-service teachers is limited in the sense that in general they are not in the actual practice of teaching. Few exceptions in research have combined the formal education process of teachers with the actual practice of teachers. One of these examples is the work of Leavy (2010), who developed a sequence for a course in a formal teacher education program to prepare teachers to teach informal inferential reasoning. The sequence followed the principles of lesson study in which pre-service teachers design, teach, and reflect about a lesson. The participants in this study demonstrated proficiency in reasoning about the elements fundamental to informal inferential reasoning but experienced difficulty in applying their knowledge to their practices. The work of Froelich et al. (2005) followed and mentored teaching assistants throughout an apprenticeship-like process while teaching an introductory statistics course. The researchers found that the close informal mentoring for graduate students was very effective in training them to be conscientious teachers of introductory statistics.

### 2.5.3 Professional Development of In-Service Teachers

Research on programs to prepare in-service teachers to teach statistics is not very common. However, the literature reports a few efforts—some sponsored by national agencies—to prepare in-service teachers to teach statistics. For example, Parsian and Rejali (2011) briefly reported on an extended program to prepare in-service statistics teachers in Iran. The program was based on a series of workshops oriented for professional statisticians. North et al. (2014) reported a massive program to prepare in-service teachers to teach statistics in South Africa. Their program invited teachers to participate in five workshops spread throughout the school year. Although the authors reported that the program was successful in improving the teachers' confidence to teach statistics, there was no evidence that the teachers implemented in their schools what they learned in the program. Massive programs to prepare in-service teachers are a great source to disseminate progress on content knowledge, technological tools, and teaching resources; however, as Ponte (2011) has stated, the most serious problem in teacher education is the low impact professional development often has on teachers' practice (c.f. Yoon et al. 2007). Massive programs always leave behind the question about the impact in the school and position the teachers as technicians—consumers of knowledge produced by others. The programs just described followed a rationalist paradigm in teacher preparation and looked at teaching as a technical and routine activity.

Other initiatives within a naturalistic perspective on teacher preparation look at teachers as learners of statistics and develop strategies to facilitate their learning. Oesterhaus and Biehler (2014) described a one-year professional development course for teaching statistical inference at the high-school level using an illustrative teaching approach called BeSt@Kontext. This teaching approach proposed extended use of computer-based simulations and dynamic visualizations in authentic contexts. Wessels (2014) described a professional development program for 8–12th grade mathematics teachers based on extensive use of hands-on activities. The results revealed growth in teachers' levels of reasoning about variability. Jacob et al. (2015) described the growth of teachers' knowledge about sampling variability in a master course for in-service teachers. Martins et al. (2015) reported the effect of a professional development program for secondary teachers aiming to increase the teachers' understanding of the relation of sample size and representativeness. The results gained through interviews with two teachers showed considerable success in this program.

### 2.5.4 A Holistic Perspective for Teacher Preparation

In recent years, research related to teacher preparation to teach statistics has shown an interesting transformation. The focus has not exclusively been on the teacher content knowledge but on other holistic forms of teacher knowledge in which the reflection of teaching experiences informs teachers' practice. Some programs for preparing in-service teachers have taken into account the teachers' contexts to build their professional knowledge. For example, Makar and Fielding-Wells (2011) worked with in-service teachers to support them in the design of inquiry-based lessons. The information collected from interviews helped the authors construct a model for understanding teachers' evolving experiences in developing expertise and confidence in learning to teach statistical inquiry. Souza et al. (2014) developed a program for early childhood education in-service teachers to support them in the teaching of stochastics. The results showed that teachers did learn something about teaching stochastics in spite of the poor statistical background with which they started. Nacarato and Grando (2014) carried out a program for in-service teachers in which a community—teachers together with university professors—prepared sequences of teaching that were further developed in teachers' classrooms. The results indicated that teachers learned from their own experience and discussion with others through the analysis of class videos. Zapata-Cardona (2014) conducted a professional development program integrating statistics teachers' experiences with formal knowledge of statistics and teaching statistics to create a community of practice. The article discussed some strengths, weaknesses, and implications for teacher preparation. Sánchez-Sánchez and Gómez-Blancarte (2015) designed a development program inspired by the methodology of the Lesson Study Group in which the teachers designed, observed, analyzed, and adjusted statistics lessons. The results described that the teachers' learning process through working together promoted their content knowledge and pedagogical content knowledge. de Souza et al. (2015) carried out a year-long program with 16 volunteer middle-school teachers in which they designed, reflected, and adjusted statistics lessons. The program showed that teachers improved their statistics teaching. This body of research has a new conception of statistics teachers, who transform teaching through reflection of their own experiences and can be positioned in the critical paradigm of teacher preparation.

### 2.5.5 Conclusion

Although research referring to teacher professional development programs is increasing in education as well as mathematics education research in recent years (e.g., Yoon et al. 2007), research in statistics education has yet gained few results in this field. Existing research can be distinguished by three paradigms as the basis of teacher preparation: the rationalist, the naturalistic, and the critical. Although teacher preparation programs based on each of these paradigms differ considerably, research has gained several recommendations for potentially efficient teacher preparation programs and professional development programs (e.g., Yoon et al. 2007). Some of these recommendations (see also Ponte 2011) are: relation to professional practice, connection with the culture of the school, challenge and support for teachers, and starting from teachers' needs and resources.

## 2.6 *Student Knowledge of Statistics*

Shaughnessy (2007) analyzed in his review a series of studies that focused on students' reasoning about certain statistical topics such as averages, variability, association, covariation, correlation, samples and surveys, and graphs. In recent years, there have been several studies that enhance the research results reported by

Shaughnessy. However, in this review, our main aim is to discuss trends in recent research on student knowledge rather than to discuss research results on the abovementioned topics in detail.

### 2.6.1 Instruments for Investigating Student Knowledge

A group of research studies has focused on developing tools and instruments to assess student knowledge in statistics. The pioneers were Garfield and delMas (2010), who, through the ARTIST project (Assessment Resource Tolls for Improving Statistical Thinking), provided a reliable, valid, practical, and accessible assessment instrument items and resources to assess students. Within that project they developed the CAOS test, which measures important aspects of statistical literacy and reasoning of college students in an introductory course. To validate the CAOS test, delMas (2014) analyzed the responses of over 30,000 secondary and tertiary level students that have taken the test over an eight-year period. The author used students' results to conduct a confirmatory factor analysis to provide evidence for the dimensionality and reliability of the instrument.

Further scholars have created other instruments or validated those already existing. For example, Baglin (2014) developed and validated an instrument to assess the development of students' statistical thinking using the framework suggested by Wild and Pfannkuch (1999). Jacobbe et al. (2014a, b) carried out the Levels of Conceptual Understanding in Statistics (LOCUS) project, which focused on assessments of statistical literacy for students in Grades 6–12. The goal was to develop assessments in accordance with the Common Core State Standards (CCSS) in the United States and with the levels hypothesized in the GAISE report (Franklin et al. 2007). Jacobbe et al. (2014a, b) carried out a pilot study assessing 1249 students' statistical understanding at the high school level (ages 14–18). The results from this study were used to refine an instrument to assess the conceptual understanding of students' school-based statistics.

### 2.6.2 Cross-Sectional Studies of Student Knowledge

Cross-sectional studies aiming to take a one-time measure of students' statistics knowledge without any intervention refer to a huge amount of different statistical topics. In several papers, Nikiforidou and Pange (2011) investigated the ability of pre-primary-level children to make sense of probabilities or to make decisions and inferences in situations of uncertainty. Based on a variety of samples the authors conclude that even young children are able to deal with risk or probabilities when faced with appropriate tasks. Eichler and Vogel (2012) also focused on young students (fourth and sixth grade). Studying a sample of 178 primary and secondary students with no previous instruction in statistics, they investigated the students' ability to understand the structure of situations of uncertainty. They found that the way of presenting a situation of uncertainty had a strong impact on the students' achievement. Yoclu and Haser (2013) analyzed 1074 eighth graders concerning their knowledge about average and variability. They found that the students were able to deal with specific concepts of averages and variability (i.e., the arithmetic mean and the range), but had difficulties in interpreting the value of a measurement. Mayén et al. (2009) studied the responses of secondary and high school students to an open-ended problem related to the concept of the median. Students had representational, conceptual, procedural, and argumentative conflicts in calculating the median and in selecting the appropriate measure to describe the center.

A huge amount of research of student knowledge refers to tertiary students (c.f. Batanero et al. 2011). For example, Hogan et al. (2015) studied how college students interpret effect size in research reports and found that students overestimated effect size. Wroughton et al. (2013) studied the role of context in college students' understanding of sampling. The researchers gave tests and interviewed students with context-based tasks. The results revealed that the students ignored the contexts in the written test but relied on context-based opinions rather than statistical principles in the interviews. Noll and Hancock (2015) interviewed 11 undergraduate students concerning their understanding of distribution and sampling. They found that the students gave two metonymies—a sort of overgeneralization when describing a mathematical concept—when discussing distributions and sampling distributions, e.g., using the features of the normal distribution as a strategy to describe further distributions. Hjalmarson et al. (2011) studied college students' responses to a model-eliciting activity in which they were asked to measure the roughness of a surface. Students' responses indicated numerous misunderstandings about center and variability. Castro-Sotos et al. (2009) surveyed 279 college students at the end of an introductory statistics course and carried out some semi-structured interviews to study correlation. The results revealed that students held strong misconceptions. Kaplan and Du (2009) explored the influence of question format in two-stage conditional probability tasks on college students' correct responses. The researchers found that the format of the question did not have any influence on student responses, but the students who used a tree diagram were more successful in calculating the conditional probability than those who did not use a tree diagram. Böcherer-Linder et al. (2015) also investigated the problem-solving abilities of students of different fields of study when presented with conditional probabilities and Bayes theorem using the visualizations of a tree diagram and of a unit square. They found that the unit square was more efficient than the tree diagram for solving tasks dealing with conditional probabilities.

### 2.6.3 Students' Knowledge Development in Intervention Designs

Another body of research has focused on students' reasoning after they have been exposed to an educational intervention. The main purpose of these studies has been to assess the impact of such interventions and to provide empirical evidence that their use can contribute to improving the teaching of statistics. For example, in one study Meletiou-Mavrotheris and Appiou-Nikiforou (2015) carried out a teaching experiment with graduate students. The participants were required to use dynamic software to create and test statistical models to solve complex real-world problems. They found that the students developed a coherent understanding of fundamental concepts of statistical inference. In another study, Saldanha (2015) developed an instructional intervention with ninth graders that involved the use of TinkerPlots for exploratory analysis of data and informal inference. The repeated simulation process allowed students to express some levels of certainty in their conclusions. Kaplan et al. (2014) designed an educational intervention to help college students to deal with the notion of randomness. Findings showed that the students in the intervention group performed better in a test about random sampling and tended to have more connections between the concepts that underlie random processes than those students in the control sample. In another study, Reaburn (2014) studied college students' understanding of $p$-values in four consecutive introductory statistics courses. The author showed that students' improvement in their understanding of $p$-values could have been due to a combination of several strategies in teaching, including the use of computer simulations and writing about mathematics. Tintle et al. (2012) used the CAOS test to compare college students' retention. One group took a randomization-based introductory statistics course, and a control group took a traditional curriculum. Students in the randomization-based group showed higher conceptual retention than those in the consensus curriculum.

### 2.6.4 Students' Thinking Processes

One of the specific approaches used by intervention studies is describing the process of an increased reasoning as a consequence of an intervention. For example Abrahamson (2012) investigated the thinking processes of primary students when analyzing the sample space of random generators that were not common for the students. He argued that experimentation could be misleading compared to a more theoretical perception of random generators. Schnell and Prediger (2012) described the process of gaining an appropriate concept of probabilities as a strategy to estimate frequencies in short series and long series of random experiments. With the data generated, the researchers built a model to explain students' reasoning about uncertainty. With a pre-test/post-test design, Meletiou-Mavrotheris and Paparistodemou (2015) investigated the impact of a teaching experiment on 69 upper-level elementary students' statistical reasoning. A pre-test including three open-ended tasks showed the students' poor statistical reasoning. The main focus of this research was to analyze the increasing statistical reasoning during the teaching experiment that was documented by audio and video recordings. The results showed that a learning trajectory that emphasizes how to select samples for a statistical investigation had an appropriate impact on the students' awareness of representativeness and of ways to ensure representativeness. Pfannkuch et al. (2015) investigated emergent reasoning about sampling variability in 11-year-old students in a learning experiment. Whereas the students were not able to describe the relation between samples and population in a pre-test, the authors showed that a teaching experiment could be sufficient to develop the students' statistical thinking about samples and sampling variability in a process. Finally, Konold et al. (2015) reported different strategies to increase students' awareness of the need to aggregate data for useful assertions about populations.

### 2.6.5 Conclusion

The majority of the studies related to students' knowledge of statistics in the last decade have been carried out with college students, some with secondary schools students, and few with elementary school students. As the mathematics curricula of many countries include some components of statistics in the first years of elementary school, this finding suggests a gap in the research. From a teacher education point of view, elementary-level research is very informative. Teacher educators and prospective teachers need to know the tensions and achievements in developing statistical knowledge of those in the very early years of elementary school. Further, there is still a strong interest in studying what students know and do not know from information gathered from multiple-choice tests, which has led to a long list of student misconceptions in statistics. Since statistical thinking is a multidimensional construct that is stimulated through problem solving (Wild and Pfannkuch 1999), research about student knowledge could focus more on the process of problem solving and thereby promote student knowledge as a result of being active participants in that process.

## 2.7 *Students' Statistics-Related Affect*

Compared to the vast amount of research concerning students' knowledge of statistics, research on students' statistics-related affect seems to be much less. For this reason, Shaughnessy (2007) has called for an increased effort to investigate students' attitudes towards and beliefs about statistics. However, there is some research referring to the statistics-related affect of learners in a broader sense. Since the biggest part of this research is based on the SATS scale (Schau et al. 1995), we do not regard beliefs, motivation, and attitudes of students in different paragraphs as we did concerning statistics teachers' affect. Rather, we first discuss the development of instruments to measure students' statistics-related affect and afterwards review research that applies the instruments and adds further research.

### 2.7.1 Instruments for Measuring Students' Statistics-Related Affect

The SATS seems to be the most applied instrument to measure students' statistics-related affect. The SATS was developed in two versions. The first version (SATS-28; Schau et al. 1995) included 28 items in categories called affect, cognitive competence, value, and difficulty, and the second version (SATS-36; Schau 2003) includes two additional categories called interest and effort. Ramirez et al. (2012) discussed the integration of the SATS into a theoretical model aiming to explain students' achievement. Nolan et al. (2012) provided a systematic statistical analysis of the characteristics of the SATS. Hood et al. (2012) discussed the basis of confirmatory factor analysis, suggesting, for example, a four-factor model as an alternative to the six-factor model.

A second instrument that is used in resent research is the Statistics Anxiety Rating Scale (STARS; c.f. Ramirez et al. 2012), which includes six categories named "worth of statistics," "interpretation anxiety," "test and class anxiety," "computation self-concept," "fear of asking for help and fear of statistics trainers." The six-factor model of STARS was confirmed by Hanna et al. (2008). Although there is a considerable number of further instruments described in the literature (Nolan et al. 2012), recent research on students' statistics-related affect is mostly based on the two instruments outlined here.

### 2.7.2 Status Quo and Development of Students' Statistics-Related Affect

A basic question of research into students' statistics-related affect is its development as a result of teaching statistics. Referring to this question and based on the SATS, Schau and Emmioglu (2012) presented a large-scale investigation of 2200 students enrolled in post-secondary introductory statistics courses. Interestingly, none of the different factors of students' statistics-related affect was significantly changed through the courses. Although in specific statistics courses there was sometimes an overall change in beliefs as a result of the courses reported (e.g., Swanson et al. 2014), Posner (2014) stated that instruments like "the Survey of Attitudes Toward Statistics (SATS) fail to show an increase in student attitudes over time (p. 1)." He argues that in most of the related research the SATS was administered shortly after the beginning of the course when the students' statistics-related affect had potentially already changed. Based on a sample of 150 students, Millar and White (2014) argued that the timing of the post-test study before final exams resulted in the phenomenon of students' unchanged affect after an introductory course in statistics.

The research of Bond et al. (2012) is slightly different from other research cited in this section. They collected individual answers to six questions referring to the nature of statistics or typical content in a statistics course. The students' statements were arranged into six categories. The results of a pre- and a post-test study in which the SATS was also administered showed students' perception of statistics had increased. Interestingly, the constructs measured with SATS did not correlate with the students' perception of statistics. A similar mixed-methods approach was used by Griffith et al. (2012). They asked 684 students whether their attitude towards statistics was positive or negative and asked further for a rationale for their answers. They categorized the students' answers into six categories of rationales for a positive or negative attitude towards statistics. The results showed that the majority of students held positive attitudes. Further, some of the students' rationales were similar to three factors in the SATS (value, difficulty, and cognitive competence), whereas no student addressed factors such as interest or effort. In contrast an often used rationale for a positive or negative attitude towards statistics was the professor of the statistics course.

One of the studies referring to students' statistics-related affect explicitly addressed students' beliefs: Muis et al. (2011) investigated the differences in students' epistemic beliefs about statistical knowledge in general and procedural knowledge and conceptual knowledge towards statistics. They investigated the beliefs of 58 students over the course of one year using the same questionnaire, partly emphasizing procedural knowledge and partly emphasizing conceptual knowledge. The result gave evidence that students' procedural knowledge and conceptual knowledge of statistics are different constructs.

One of the few studies focusing on students in schools is provided by Huynh et al. (2014). In a pilot study based on a three-item questionnaire they found a positive impact of a learning experiment on students' attitudes. A further research approach is provided by Eichler (2008). Using interviews, he investigated the knowledge and beliefs (values) towards statistics of 20 upper-secondary students. He found that the degree to which students assign statistics a value for personal life and society is dependent on their teachers' beliefs.

### 2.7.3 Comparison of Different Populations and Courses

An important aspect of existing research has been the investigation of the impact of both course characteristics and student characteristics on students' statistics-related affect. Several research approaches in this field were based on the SATS. For example, Posner (2011) compared the impact of a course with a new assessment system to a course with traditional assessments. In two groups each with 30 students, Posner reported significantly different student affect on all subscales of the SATS. For example, students with the new assessment assigned statistics a bigger value than students in the course with traditional assessments. DeVaney (2010) compared the affect of students enrolled in an online course and in a traditional on-campus course. He found that the students' affect in the online course was significantly higher on the attitude scale (e.g., "I will like statistics.") and was significantly lower concerning the perceived difficulty of statistics than in the traditional course. Further, Gundlach et al. (2015) compared three courses: a traditional course, an online course, and a course applying a flipped classroom method. Measures of the students' related affect showed that the increase on different subscales of the SATS was significantly higher in the traditional course than in the other two courses. In contrast, Ramirez and Bond (2014) found no differences between students' affect in a traditional course with online elements and in a project-based course. Similarly, Swanson et al. (2014) did not find the mentioned differences of students enrolled in a traditional course and in a course that emphasized simulation in terms of randomization, permutation tests, or bootstrapping.

Williams (2010) researched teachers' impact on students' statistics-related affect using the STARS. She compared two groups of 76 students from different fields of study. The first group was taught by the researcher with an approach that emphasized immediacy. Another instructor taught the same syllabus without emphasizing immediacy. Williams reported that on the one hand the students were aware of the instructors' differences in emphasizing immediacy and that on the other hand the different groups were significantly different on the STARS scales.

A further aspect of a comparison is to distinguish two populations in terms of their statistics-related affect. For example, Chiesi and Primi (2015) reported that gender has an impact on students' statistics-related affect. Investigating 177 students, they found that female students hold more negative attitudes towards statistics and rate their statistical competence lower than male students although they did not differ in their achievement. DeVaney (2010) showed that students felt less anxiety when enrolled in an online course whereas the anxiety in a traditional course persisted.

### 2.7.4 Relation of Statistics-Related Affect to Further Variables

There is a wide consensus that the relation of dispositional elements and knowledge elements are the most important relations between statistics-related affect and further variables. For this line of research, Emmioglu and Capa-Aydin (2012) provided a meta-analysis of 17 studies from 1998 to 2011 that addressed the relation between students' statistics-related affect measured with the SATS and students' achievement. The results of the meta-analysis suggest a permanent significant correlation between subscales of the SATS—beliefs, motivation, and attitudes—and students' achievement. Chiesi and Primi (2010), who conducted one of the studies included in the aforementioned meta-analysis, developed a linear model of cognitive and non-cognitive variables that have an impact on students' achievement. They found that mathematical knowledge has the biggest impact on students' achievement in a statistics course, whereas students' statistics-related affect has a small impact and anxiety (measured with the STARS) has no impact. Also Hood et al. (2012) reported past performance in statistics to be the best predictor of students' future achievement, whereas variables included in SATS-28 are at most indirectly connected with students' achievement. Budé et al. (2007) developed a linear model that described the impact of motivational variables on students' achievement. Parts of the items from different variables were translated from SATS, finding the attitudes scale and the persistence scale to have a strong impact on the students' achievement. Vanhoof et al. (2006) conducted a longitudinal study in which they measured students' statistics-related affect using the ATS (Wise 1985) and results in exams. They found a significant correlation between students' affect and students' achievement for the first year; however, in the second year only part of students' affect correlates significantly to the students' achievement and there was no significant correlation in the following years. Interestingly, the authors found the students' statistics-related affect concerning their field of study to have a higher correlation with their dissertation grade than their achievement in the first-year exam. The STARS-based longitudinal research of Keeley et al. (2008) suggested the result that the correlation of subscales of the STARS and the students' final achievement gets higher and higher during the course, whereas the anxiety decreases during the course.

Besides research focusing on the relation between affect and achievement, there were some other relations studied. For example, using two scales referring to "worry" and "intolerance of uncertainty" together with the STARS, Williams (2013) found a strong correlation between the three constructs. A fundamentally different research approach was reported Neumann et al. (2013). They conducted a course using real-life data and investigated through interviews which topics students relate to the use of real-life data. They found motivational aspects such as interest and also students' statements that real-life data fosters understanding and remembering.

### 2.7.5 Conclusion

One aspect is striking when reflecting on existing research on students' statistics-related affect: Statistics education research has gained a great deal of knowledge referring to students of different fields of study at the university level. In contrast, research has gained—with a few exceptions—practically no knowledge about students statistics-related affect. For this reason, we can repeat part of Shaughnessy's (2007) call: For future research it is crucial to put more effort into research on school students' beliefs, motivation, and attitudes towards statistics.

An important result of research is that both parts of statistical literacy, i.e., knowledge elements and dispositional elements, have an impact on students' achievement in statistics, as the different developments of linear models imply (e.g., Chiesi and Primi 2010). Although these models also imply that the most important factor is the mathematical or statistical knowledge before a course, the research of Vanhoof et al. (2006) gave evidence that students' statistics-related affect is a main predictor of students' achievement in the long term.

Finally, discrepancies among the results of different studies that seem to be similar are striking. Whereas on the one hand a certain development of factors of students' statistics-related affect is reported, other research suggests the independence of students' affect and the teaching of statistics. A potential reason for the messiness of these results is the lack of a context relation of the research. Thus, in many research reports the profession of the investigated students, the teacher who taught in the courses, and the teaching orientation in the classroom are not regarded as a main factor. For example, why should psychology students value statistics in the same way as economic students do, or is a traditional way of teaching in one country or university the same as in another country or university? Another aspect for future research that is also related to a specific context might be to focus more on parts of students' affect in relation to the aims of a statistics course. Thus, it might be a difference to change beliefs that are understood to be relatively stable (e.g., Philipp 2007) in comparison to attitudes in a short-term statistics course.

## 2.8 Technology as Facilitator of Statistics Learning

Although there is research that has focused on teachers and technology, we have included related research in this section. We proceed in this way because most of the research focusing on teachers and technology regards teachers as learners without a specific classroom context. We further connect the following research to the review of Biehler et al. (2013). In this review, different topics of research were identified: research that focuses on the comparison of distributions, on a statistical investigation in general, on informal statistical inference, on the relation of data and chance, and, finally, on pathways to statistical inference. In the sources that we used for our review we did not find research reports that primarily focused on the last two fields of research.

Referring to the comparison of distributions, Madden (2014) analyzed interviews with pre-service and in-service teachers that participated in a course aiming to enhance the teachers' knowledge about comparing distributions supported by Fathom. She concluded that the use of technology in particular resulted in the teachers' increased statistical reasoning concerning distributions. Frischemeier and Biehler (2015) also reported the development of an acceptable ability of pre-service teachers to compare different groups based on a course using TinkerPlots.

The research of Ben-Zvi et al. (2012) is related to statistical investigations in general. Their research aimed to analyze the growth of students' perception of the influence of sample size as part of a statistical investigation. They documented the process of students' learning using transcribed video observations. The results of this research showed that using TinkerPlots enabled the students to develop their statistics-related language. Thus in the beginning, the students used mostly arguments based on "certainty-only (deterministic) and uncertainty-only (relativistic) statements" (p. 913), whereas the students developed a more sophisticated probabilistic language through the course. Referring to the same research program, Manor-Braham and Ben-Zvi (2015) used TinkerPlots to explore students' use of sample distributions and informal inferential reasoning. Burgess (2014) focused on the primary students' acceptance of the use of TinkerPlots in a statistical investigation. Based on interviews and video recordings, he reported in particular the students' beliefs that technology facilitates the investigation. In a way similar to the research that we looked at in the section concerning students' statistics related affect, Garfield et al. (2012) investigated the impact on the students' knowledge and attitudes of an introductory statistics course named CATALST that was based on the use of TinkerPlots. They compared the test achievement of the students in their study to the achievement of students at other universities based on the assumption that the latter students were enrolled in traditional courses. Although the authors reported a considerable growth in the students' statistical reasoning ability in their course using TinkerPlots, they stated that "in general, there seems to be little difference between the CATALST and non-CATALST students on most of the items" (Garfield et al. 2012, p. 895).

Lee and Nickell (2014) analyzed the growth of pre-service teachers' knowledge about a statistical investigation supported by the use of Fathom. They reported increased statistical knowledge as well as a better understanding of using technology to teach statistics. Similar to the approach of Lee and Nickell (2014), Meletiou-Mavrotheris et al. (2014) described the learning of pre-service teachers that used TinkerPlots to examine sampling as a part of informal inferential reasoning. The results indicate that by using TinkerPlots the pre-service teachers enhanced their knowledge about sampling and also developed ideas to integrate sampling into their statistics teaching. Looking at courses for teachers at eight U.S. institutions, Lee et al. (2014) analyzed the growth of 62 teachers' statistical knowledge with a specific focus on these teachers' ability to represent data.

A specific form of using technology, the use of microworlds, is included in the research of Pratt et al. (2012). This approach involves computer-generated environments that allow students to analyze data, make experiments, or investigate statistical models in a virtual world. In this research, the microworld included a situation (Deborah's dilemma) in which pairs of teachers could manipulate parameters that have an impact on Deborah's pain level, which depends on the decision to have a (reasonable) operation or not. One main result of the teachers' decision-making in risk situations was that the decisions were based on a prior heuristic, i.e., making decisions without considering the probability of this impact.

### 2.8.1 Conclusion

Research concerning technology focuses on the two main aspects for which technology could serve as a facilitator of learning mathematics and statistics (cf. Doerr and Zangor 2000). The first aspect includes supporting statistical investigation and, thus, statistical reasoning. The second aspect includes supporting conceptual learning, e.g., through simulation. In these fields of research it is striking that technology seems to be restricted to software such as TinkerPlots or Fathom. Actually, there are reports of technology-supported statistics courses that are based on other software. However, only a few approaches refer to empirical research. For example, Xu et al. (2014) referred to the acceptance of specific software such as SPSS in a course on biostatistics and Ferreira et al. (2014) reported the efficiency of R to support the high school students' learning. By contrast, reports focusing on software but not focusing on TinkerPlots or Fathom mostly refer to curriculum development.

Research referring to students' learning of statistics using technology has gained a lot of knowledge concerning pathways to different statistical concepts and the role that technology could play in respect to these pathways. By contrast, research in this field has not provided sufficient statistical evidence showing the efficiency of teaching statistics with technology.

## 3 Conclusion

We firstly refer to the introduction of this paper: The research boom in statistics education that Shaughnessy (2007) a decade ago is still alive. The amount of papers that refer to empirical research into teachers' and students' statistical knowledge and statistics-related affect seems to be overwhelming in some areas. However, there are other areas in statistics education research that need special attention. In this concluding section we discuss some of research gaps or shortcomings that we identified in our literature review.

### *3.1 A Missing Norm for Statistical Knowledge*

The research into teachers' knowledge has often led to the conclusion that teachers were not able to interpret statistical concepts in an appropriate way or that teachers hold a poor knowledge towards specific statistical topics. However, we are not aware of a definition for a norm for teachers' appropriate knowledge. Some countries' standards for a teacher preparation include a consensus about appropriate mathematical knowledge and mathematical reasoning. This consideration is not in the same sense relevant with respect to students' statistical knowledge since there is often a concrete norm that is represented, for example, by national standards.

Although a norm could potentially improve the interpretation of deficits in teachers' statistical knowledge, a further perspective on teachers' knowledge could be to describe the teachers' existing knowledge: Although it is important to determine what teachers do not know, it is also important to find out what they do know. Research that investigates teachers as experts of statistics teaching is scarce. By contrast, a considerable number of research studies referring to students' statistical knowledge address the students' existing or increasing knowledge (e.g., Schnell and Prediger 2012, or Manor-Braham and Ben-Zvi 2015).

### *3.2 Content-Relatedness*

First we refer to teachers' statistical knowledge again to outline what we identified as a research gap: Whereas we gained a considerable amount of information about teachers' knowledge in general, we gained little knowledge about teachers' context-related knowledge. More research on teachers' knowledge from the perspective of these teachers' practice could greatly enhance our knowledge towards teachers' statistical knowledge.

A further aspect of research on teachers' statistical knowledge in terms of content-relatedness is the status of the teachers being studied: Is it more appropriate to allocate the knowledge of pre-service teachers to teachers' or learners' knowledge? Although it is often easier to recruit a sample of pre-service teachers than in-service teachers and research with pre-service teachers' has gained important results, research with pre-service teachers is not able to yield information about these teachers' knowledge in relation to their professional practice.

A stronger focus on the context of teaching practice could potentially also enhance our knowledge of

- aspects of teachers' pedagogical content knowledge, which according to Shulman (1986) represents the amalgam of content knowledge and a content-specific pedagogical knowledge. Research focusing on teachers' pedagogical content knowledge is underrepresented in the body of research into statistics teachers (c.f. Chick and Pierce 2011).
- the teachers' classroom practice. As mentioned above, research on this topic is scarce (Eichler 2011). However the teachers' classroom practice could provide a deeper insight into teachers' knowledge than a study of teachers' pedagogical content knowledge that is detached from a specific classroom practice (c.f. Ponte 2011).
- the relation between the teachers' knowledge and statistics-related affect and their students' knowledge and statistics-related affect.

Finally, teacher preparation could be examined in a context-related way by applying a holistic perspective to it that takes into account teachers' specific practice in such programs. In general, research into teacher professional development is underrepresented in statistics education research.

## *3.3 The Need for Research on Students' Statistics-Related Affect*

Different research approaches suggest that students' statistics-related affect is an important factor in students' achievement (e.g., Chiesi and Primi 2010) and could even be understood as the most important factor in students' achievement in the long run (Vanhoof et al. 2006). Although the latter research result waits for replication or confirmation on the basis of similar research results, it suggests intensifying research into students' statistics-related affect.

Existing research in this field, however, refers nearly without exception to tertiary students. For this reason there is an ongoing need to investigate elementary, middle and high school students' attitudes, beliefs, and motivation. This research becomes more important the more students' affect has an impact on students' achievement and also students' later engagement in statistics.

For both research into students' and teachers' statistics-related affect it may be reasonable to reconsider the investigated constructs, i.e., stable dispositions such as beliefs, more changeable dispositions such as attitudes, or an individual's motivation in terms of interest or self-concept.

## 3.4 *The Best Method for Research in Statistics Education*

Currently, there is a certain consistency in the methods used to research topics. For example, research into students' statistics-related affect follows mostly a quantitative method. In order to get a different perspective, it would be worthwhile to put additional effort into also investigating students' statistics-related affect using qualitative methods to reveal, for example, systems of beliefs or attitudes. By contrast, research addressing technology as a facilitator of student learning is mostly based on qualitative methods that aim to describe the process of student learning in detail. However, it may also be worthwhile to apply quantitative methods in this field, for example, to compare statistically the efficiency of learning with and without technology.

In general, a wish for future research into statistics education is to approach each of the six research topics that we addressed with a variety of methods or even mixed methods within one study. However, independent from the method, an ongoing aim for research in statistics education is to maintain the boom in this line of research.

### Further Reading

This survey on the state of the art of research in statistics education was explicitly restricted to empirical research. Further reading is available with findings on these topics:

- Approaches of curriculum development. For example, some papers in Batanero et al. (2011) addressed this aspect of statistics education research.
- Models of statistical thinking (Wild and Pfannkuch 1999), statistical reasoning (Garfield 2002), or statistical literacy (Gal 2002). In this survey, we only examine the latter model.
- Elaborations of a statistics curriculum (e.g., Garfield and Ben-Zvi 2008).

In addition, since most of the empirical research we examined was published after existing reviews of the status quo in statistics education research (e.g., Ben-Zvi and Garfield 2004; Jones et al. 2007; Shaughnessy 2007; Batanero et al. 2011), a crucial approach to prior research in statistics education is given in the abovementioned reviews.

# References

Abrahamson, D. (2012). Seeing chance: Perceptual reasoning as an epistemic resource for grounding compound event spaces. *ZDM Mathematics Education, 44*(7), 869–881.

Arnold, P. (2008). Developing new statistical content knowledge with secondary school mathematics teachers. In C. Batanero, G. Burrill, C. Reading, & A. Rossman (Eds.), *Joint ICMI/IASE Study: Teaching Statistics in School Mathematics. Challenges for Teaching and Teacher Education. Proceedings of the ICMI Study 18 and 2008 IASE Round Table Conference*. Monterey: México.

Baglin, J. (2014). Discerning students' statistical thinking: A researcher's perspective. In K. Makar, B. de Sousa, & R. Gould (Eds.), *Sustainability in Statistics Education. Proceedings of the Ninth International Conference on Teaching Statistics (ICOTS9), Flagstaff, Arizona, USA.* Voorburg: International Association of Statistics Education.

Bakogianni, D. (2015). Studying the process of transforming a statistical inquiry-based task in the context of a teacher study group. In K. Krainer & N. Vondrova (Eds.), *Proceedings of the 9th Conference of the European Society for Research in Mathematics Education (CERME9)*. Prague, Czech Republic: Charles University.

Bandura, A. (2012). *Self-efficacy: The exercise of control* (13. printing). New York, NY: Freeman.

Bansilal, S. (2014). Using an APOS framework to understand teachers' responses to questions on the normal distribution. *Statistics Education Research Journal, 13*(2), 42–57.

Batanero, C., Burrill, G., & Reading, C. (2011). In *New ICMI study series: Vol. 14. Teaching statistics in school mathematics-challenges for teaching and teacher education: A joint ICMI/IASE study: The 18th ICMI study*. Dordrecht: Springer Science+Business Media B.V.

Batanero, C., Estrada, A., Diaz, C., & Fortuny, J. M. (2005). A structural study of future teachers' attitudes towards statistics. In M. Bosch (Ed.), *Proceedings of the Fourth Congress of the European Society for Research in Mathematics Education* (pp. 508–517). Sant Feliu de Guíxols.

Begg, A., & Edwards, R. (1999). *Teachers' ideas about teaching statistics*. Australian Association for Research in Education & New Zealand Association for Research in Education, Melbourne. Retrieved from Online: www.aare.edu.au/99pap/

Ben-Zvi, D., Aridor, K., Makar, K., & Bakker, A. (2012). Students' emergent articulations of uncertainty while making informal statistical inferences. *ZDM Mathematics Education, 44*(7), 913–925.

Ben-Zvi, D., & Garfield, J. (Eds.). (2004). *The challenge of developing statistical literacy, reasoning and thinking*. Dordrecht: Springer Science+Business Media Inc.

Biehler, R., Ben-Zvi, D., Bakker, A., & Makar, K. (2013). Technology for enhancing statistical reasoning at the school level. In M. Clements, A. J. Bishop, C. Keitel, J. Kilpatrick, & F. K. Leung (Eds.), *Third international handbook of mathematics education* (pp. 643–689). New York, NY: Springer New York.

Böcherer-Linder, K., Eichler, A., & Vogel, M. (2015). Understanding conditional probability through visualization. In H. Oliveira, A. Henriques, A. P. Canavarro, C. Monteiro, C. Carvalho, J. P. Ponte, & S. Colaço (Eds.), *Proceedings of the International Conference Turning data into knowledge: New opportunities for statistics education* (pp. 14–23). Lisbon, Portugal: Instituto de Educação da Universidade de Lisboa.

Bond, M. E., Perkins, S. N., & Ramirez, C. (2012). Students' perceptions of statistics: An exploration of attitudes, conceptualizations, and content knowledge of statistics. *Statistics Education Research Journal, 11*(1), 6–25.

Budé, L., Van de Wiel, M. W., Imbos, T., Candel, M., J., Broers, N., & Berger, M. P. (2007). Students' achievements in a statistics course in relation to motivational aspects and study behaviour. *Statistics Education Research Journal*, *6*(1), 5–21.

Burgess, T. (2011). Teaching statistics in school mathematics—Challenges for teaching and teacher education. In C. Batanero, G. Burrill, & C. Reading (Eds.), *New ICMI study series: Vol. 14. Teaching statistics in school mathematics-challenges for teaching and teacher education: A joint ICMI/IASE study: The 18th ICMI study* (pp. 259–270). Dordrecht: Springer Science+Business Media B.V.

Burgess, T. (2014). Student perspectives on being introduced to using Tinkerplots for investigations. In K. Makar, B. de Sousa, & R. Gould (Eds.), *Sustainability in Statistics Education. Proceedings of the Ninth International Conference on Teaching Statistics (ICOTS9), Flagstaff, Arizona, USA.* Voorburg: International Association of Statistics Education.

Casey, S. A. (2010). Subject matter knowledge for teaching statistical association. *Statistics Education Research Journal, 9*(2), 50–68.

Casey, S. A., & Wasserman, N. H. (2015). Teachers' knowledge about informal line of best fit. *Statistics Education Research Journal, 14*(1), 8–35.

Castro-Sotos, A. E., Vanhoof, S., Van Den Noortgate, W., & Onghena, P. (2009). The transitivity misconception of Pearson's correlation coefficient. *Statistics Education Research Journal, 8*(2), 33–55.

Chick, H., & Pierce, R. (2011). Teachers' beliefs about statistics education. In C. Batanero, G. Burrill, & C. Reading (Eds.), *New ICMI study series: Vol. 14. Teaching statistics in school mathematics-challenges for teaching and teacher education: A joint ICMI/IASE study: The 18th ICMI Study* (pp. 151–162). Dordrecht: Springer Science+Business Media B.V.

Chiesi, F., & Primi, C. (2010). Cognitive and non-cognitive factors related to students' statistics achievement. *Statistics Education Research Journal, 9*(1), 6–26.

Chiesi, F., & Primi, C. (2015). Gender differences in attitudes toward statistics: Is there a case for a confidence gap? In K. Krainer & N. Vondrova (Eds.), *Proceedings of the 9th Conference of the European Society for Research in Mathematics Education (CERME9).* Prague, Czech Republic: Charles University.

Cobb, G. W., & Moore, D. S. (1997). Mathematics, statistics, and teaching. *American Mathematical Monthly, 104*, 801–823.

da Silva, C., Kataoka, V., & Cazorla, I. (2014). Analysis of teachers' understanding of variation in the dot-boxplot context. In K. Makar, B. de Sousa, & R. Gould (Eds.), *Sustainability in Statistics Education. Proceedings of the Ninth International Conference on Teaching Statistics (ICOTS9), Flagstaff, Arizona, USA.* Voorburg: International Association of Statistics Education.

de Souza, L., Lopes, C. E., & Pfannkuch, M. (2015). Collaborative professional development for statistics teaching: a case study of two middle-school mathematics teachers. *Statistics Education Research Journal, 14*(1), 112–134.

delMas, R. (2014). Trends in students' conceptual understanding of statistics. In K. Makar, B. de Sousa, & R. Gould (Eds.), *Sustainability in Statistics Education. Proceedings of the Ninth International Conference on Teaching Statistics (ICOTS9), Flagstaff, Arizona, USA.* Voorburg: International Association of Statistics Education.

DeVaney, T. A. (2010). Anxiety and attitude of graduate students in on-campus vs. online statistics courses. *Journal of Statistics Education, 18*(1).

Di Martino, P., & Zan, R. (2010). 'Me and maths': Towards a definition of attitude grounded on students' narratives. *Journal of Mathematics Teacher Education, 13*(1), 27–48.

Doerr, H. M., & Zangor, R. (2000). Creating meaning for and with the graphic calculator. *Educational Studies in Mathematics, 41*(2), 143–163.

Dolor, J., & Noll, J. (2015). Using guided reinvention to develop teachers' understanding of hypothesis testing concepts. *Statistics Education Research Journal, 14*(1), 60–89.

Eagly, A. H., & Chaiken, S. (1998). Attitude structure and function. In D. T. Gilbert, S. T. Fiske, & G. Lindzey (Eds.), *The handbook of social psychology* (4th ed., pp. 269–322). Boston, New York: Oxford University Press.

Eichler, A. (2008). Teachers' classroom practice in statistics courses and students' learning. In C. Batanero, G. Burrill, C. Reading, & A. Rossman (Eds.), *Joint ICMI/IASE Study: Teaching Statistics in School Mathematics. Challenges for Teaching and Teacher Education. Proceedings of the ICMI Study 18 and 2008 IASE Round Table Conference*. Monterey: México.

Eichler, A. (2011). Statistics teachers and classroom practices. In C. Batanero, G. Burrill, & C. Reading (Eds.), *New ICMI study series: Vol. 14. Teaching statistics in School mathematics-challenges for teaching and teacher education: A joint ICMI/IASE study: The 18th ICMI study* (pp. 175–186). Dordrecht: Springer Science+Business Media B.V.

Eichler, A., & Erens, R. (2015). Domain-specific belief systems of secondary mathematics teachers. In B. Pepin & B. Roesken-Winter (Eds.), *From beliefs to dynamic affect systems in mathematics education* (pp. 179–200). Cham: Springer International Publishing.

Eichler, A., & Vogel, M. (2012). Basic modelling of uncertainty: Young students' mental models. *ZDM Mathematics Education, 44*(7), 841–854.

Emmioglu, E., & Capa-Aydin, Y. (2012). Attitudes and achievement in statistics: A meta-analysis study. *Statistics Education Research Journal, 11*(2), 95–102.

Estrada, A., Batanero, C., & Lancester, S. (2011). Teachers' attitudes towards statistics. In C. Batanero, G. Burrill, & C. Reading (Eds.), *New ICMI study series: Vol. 14. Teaching statistics in school mathematics-challenges for teaching and teacher education: A joint ICMI/IASE study: The 18th ICMI study* (pp. 163–172). Dordrecht: Springer Science+Business Media B.V.

Ferreira, R. D. S., Kataoka, V. Y., & Karrer, M. (2014). Teaching probability with the support of the R statistical software. *Statistics Education Research Journal, 13*(2), 132–147.

Fives, H., & Buehl, M. M. (2012). Spring cleaning for the "messy" construct of teachers' beliefs: What are they? Which have been examined? What can they tell us? In K. R. Harris, S. Graham, & T. C. Urdan (Eds.), *APA handbooks in psychology. APA educational psychology handbook* (Vol. 2, pp. 471–499). Washington, DC: American Psychological Association.

Franklin, C., Kader, G., Mewborn, D., Moreno, J., Peck, R., & Perry, M. (2007). *Guidelines for assessment and instruction in statistics education (GAISE) report: A pre-K-12 curriculum framework*. Alexandria, VA: American Statistical Association.

Frischemeier, D., & Biehler, R. (2015). Pre-service teachers' statistical reasoning when comparing groups facilitated by software. In K. Krainer, N. Vondrova, & J. Novotna (Eds.), *Proceedings of the 9th Conference of the European Society for Research in Mathematics Education (CERME9)*. Prague, Czech Republic: Charles University.

Froelich, A. G., Duckworth, W. M., & Stephenson, W. R. (2005). Training statistics teachers at Iowa State University. *The American Statistician, 59*(1), 8–10.

Gal, I. (2002). Adults' statistical literacy: Meanings, components, responsibilities. *International Statistical Review, 70*(1), 1–25.

Garfield, J. (2002). The challenge of developing statistical reasoning. *Journal of Statistics Education, 10*(3).

Garfield, J., & Ben-Zvi, D. (2008). *Developing students' statistical reasoning: Connecting research and teaching practice*. Dordrecht: Springer.

Garfield, J., & delMas, R. (2010). A web site that provides resources for assessing students' statistical literacy, reasoning and thinking. *Teaching Statistics, 3*(1), 2–7.

Garfield, J., delMas, R., & Zieffler, A. (2012). Developing statistical modelers and thinkers in an introductory, tertiary-level statistics course. *ZDM Mathematics Education, 44*(7), 883–898.

Griffith, J. D., Adams, L. T., Gu, L. L., Hart, C. L., & Nichols-Whitehead, P. (2012). Students' attitudes toward statistics across the disciplines: A mixed-methods approach. *Statistics Education Research Journal, 11*(2), 45–56.

Groth, R. (2007). Towards a conceptualization of statistical knowledge for teaching. *Journal for Research in Mathematics Education, 38*(5), 427–437.

Groth, R. (2013). Characterizing key developmental understandings and pedagogically powerful ideas within a statistical knowledge for teaching framework. *Mathematical Thinking and Learning, 15*, 121–145.

Gundlach, E., Richards, A. R., Nelson, D., & Levesque-Bristol, C. (2015). A comparison of student attitudes, statistical reasoning, performance, and perception for web-augmented traditional, fully online, and flipped sections of s statistical literacy class. *Journal of Statistics Education, 23*(1).

Hanna, D., Shevlin, M., & Dempster, M. (2008). The structure of the statistics anxiety rating scale: A confirmatory factor analysis using UK psychology students. *Personality and Individual Differences, 45*(1), 68–74.

Hannigan, A., Gill, O., & Leavy, A. M. (2013). An investigation of prospective secondary mathematics teachers' conceptual knowledge of and attitudes towards statistics. *Journal of Mathematics Teacher Education, 16*(6), 427–449.

Hannula, M. S. (2006). Motivation in mathematics: Goals reflected in emotions. *Educational Studies in Mathematics, 63*(2), 165–178.

Hannula, M. S. (2012). Exploring new dimensions of mathematics-related affect: Embodied and social theories. *Research in Mathematics Education, 14*(2), 137–161.

Hjalmarson, M. A., Moore, T. J., & delMas, R. (2011). Statistical analysis when the data is an image: eliciting student thinking about sampling and variability. *Statistics Education Research Journal, 10*(1), 15–34.

Hobden, S. (2014). When statistical literacy really matters: Understanding published information about the HIV/AIDS epidemic in South Africa. *Statistics Education Research Journal, 13*(2), 72–82.

Hogan, T. P., Zaboski, B. A., & Perry, T. R. (2015). College students' interpretation of research reports on group differences: the tall-tale effect. *Statistics Education Research Journal, 14*(1), 90–111.

Hood, M., Creed, P. A., & Neumann, D. L. (2012). Using the expectancy value model of motivation to understand the relationship between student attitudes and achievement in statistics. *Statistics Education Research Journal, 11*(2), 72–85.

Huynh, M., Bargin, J., & Bedford, A. (2014). Improving the attitudes of high school students towards statistics: An island-based approach. In K. Makar, B. de Sousa, & R. Gould (Eds.), *Sustainability in Statistics Education. Proceedings of the Ninth International Conference on Teaching Statistics (ICOTS9), Flagstaff, Arizona, USA.* Voorburg: International Association of Statistics Education.

Jacob, B. L., Lee, H. S., Tran, D., & Doerr, H. M. (2015). Improving teachers' reasoning about sampling variability: A cross institutional effort. In K. Krainer & N. Vondrova (Eds.), *Proceedings of the 9th Conference of the European Society for Research in Mathematics Education (CERME9).* Prague, Czech Republic: Charles University.

Jacobbe, T. (2012). Elementary school teachers' understanding of the mean and median. *International Journal of Science and Mathematics Education, 10*, 1143–1161.

Jacobbe, T., Foti, S., Case, C., & Whitaker, D. (2014). High school (ages 14–18) students' understanding of statistics. In K. Makar, B. de Sousa, & R. Gould (Eds.), *Sustainability in Statistics Education. Proceedings of the Ninth International Conference on Teaching Statistics (ICOTS9), Flagstaff, Arizona, USA.* Voorburg: International Association of Statistics Education.

Jacobbe, T., & Horton, R. M. (2010). Elementary school teachers' comprehension of data displays. *Statistics Education Research Journal, 9*(1), 27–45.

Jacobbe, T., Whitaker, D., Case, C., & Foti, S. (2014). The LOCUS assessment at the college level: conceptual understanding in introductory statistics. In K. Makar, B. de Sousa, & R. Gould (Eds.), *Sustainability in Statistics Education. Proceedings of the Ninth International Conference on Teaching Statistics (ICOTS9), Flagstaff, Arizona, USA.* Voorburg: International Association of Statistics Education.

Jones, G. A., Langrall, C. W., & Mooney, E. S. (2007). Research in probability. In F. K. Lester (Ed.), *Second handbook of research on mathematics teaching and learning* (pp. 909–956). Charlotte, NC: Information Age.

Kaplan, J. J., & Du, J. (2009). Question format and representations: Do heuristics and biases apply to statistics students? *Statistics Education Research Journal, 8*(2), 33–55.

Kaplan, J. J., Rogness, N. T., & Fisher, D. G. (2014). Exploiting lexical ambiguity to help students understand the meaning of random. *Statistics Education Research Journal, 13*(1), 9–24.

Kataoka, V., da Silva, C., & Cazorla, I. (2014). Analysis of teachers' understanding of covariation in the Vitruvian man context. In K. Makar, B. de Sousa, & R. Gould (Eds.). *Sustainability in Statistics Education. Proceedings of the Ninth International Conference on Teaching Statistics (ICOTS9), Flagstaff, Arizona, USA.* Voorburg: International Association of Statistics Education.

Keeley, J., Zayac, R., & Correira, C. (2008). Curvilinear relationships between statistics anxiety and performance among undergraduate students: Evidence for optimal anxiety. *Statistics Education Research Journal, 7*(1), 4–15.

Koleza, E., & Kontogianni, A. (2013). Assessing statistical literacy: What do freshmen know? In B. Ubuz, C. Haser, & M. A. Mariotti (Eds.), *Proceedings of the Eighth Congress of the European Society for Research in Mathematics Education (CERME8).* Ankara: Middle East Technical University.

Konold, C., Higgins, T., Russell, S. J., & Khalil, K. (2015). Data seen through different lenses. *Educational Studies in Mathematics, 88*(3), 305–325.

Leavy, A. M. (2006). Using data comparison to support a focus on distribution: Examining pre-service teachers' understandings of distribution when engaged in statistical inquiry. *Statistics Education Research Journal, 5*(2), 89–114.

Leavy, A. M. (2010). The challenge of preparing pre-service teachers to teach informal inferential reasoning. *Statistics Education Research Journal, 9*(1), 46–67.

Leavy, A. M., Hannigan, A., & Fitzmaurice, O. (2013). If you're doubting yourself then, what's the fun in that? An exploration of why prospective secondary mathematics teachers perceive statistics as difficult. *Journal of Statistics Education, 21*(3).

Lee, H. S., Kersiant, G., Harper, S. R., Driskell, S., Jones, D. L., Leatham, K. R., et al. (2014). Teachers' use of transnumeration in solving statistical tasks with dynamic software. *Statistics Education Research Journal, 13*(1), 25–52.

Lee, H. S., & Nickell, J. (2014). How curriculum may develop technological statistical knowledge: A case of teachers examining relationships among variables using fathom. In K. Makar, B. de Sousa, & R. Gould (Eds.), *Sustainability in Statistics Education. Proceedings of the Ninth International Conference on Teaching Statistics (ICOTS9), Flagstaff, Arizona, USA.* Voorburg: International Association of Statistics Education.

Loya-Chávez, H. (2008). Modelos pedagógicos en la formación de profesores [Pedagogical models in teachers' preparation]. *Synthesis, 42*, 1–6.

Madden, S. (2014). Designing technology-rich learning environments for secondary teachers to explore and prepare to teach statistics. In K. Makar, B. de Sousa, & R. Gould (Eds.), *Sustainability in Statistics Education. Proceedings of the Ninth International Conference on Teaching Statistics (ICOTS9), Flagstaff, Arizona, USA.* Voorburg: International Association of Statistics Education.

Magalhães, M. N., & Magalhães, M. C. (2014). A critical understanding and transformation of an introductory statistics course. *Statistics Education Research Journal, 13*(2), 28–41.

Makar, K., & Fielding-Wells, J. (2011). Teaching teachers to teach statistical investigations. In C. Batanero, G. Burrill & C. Reading (Eds.), *New ICMI study series: Vol. 14. Teaching statistics in school mathematics-challenges for teaching and teacher education: A joint ICMI/IASE study: The 18th ICMI study* (pp. 347–358). Dordrecht: Springer Science+Business Media B.V.

Manor-Braham, H., & Ben-Zvi, D. (2015). Students' articulations of uncertainty in informally exploring sampling distributions. In A. Zieffler & E. Fry (Eds.), *Reasoning about uncertainty: Learning and teaching informal inferential reasoning.* Minneapolis: Catalyst Press.

Mathematical Science Education Board & National Research council (Ed.). (1990). *Reshaping school mathematics. A philosophy and framework for curriculum.* Washington: National Academy Press.

Martins, J. A., Nasciemento, M. M., & Estrada, A. (2012). Looking back over their shoulders: A qualitative analysis of Portuguese teachers' attitudes towards statistics. *Statistics Education Research Journal*, 11(2), 26–44.

Martins, M. N., Monteiro, C., & Carvalho, C. (2015). Analyzing the concepts of teacher about sampling using TinkerPlots 2.0. In K. Krainer, N. Vondrova, & J. Novotna (Eds.), *Proceedings of the 9th Conference of the European Society for Research in Mathematics Education (CERME9).* Prague, Czech Republic: Charles University.

Mayén, S., Díaz, C., & Batanero, C. (2009). Conflictos semióticos de estudiantes con el concepto de mediana [Students' semiotic conflicts in the concept of median]. *Statistics Education Research Journal, 8*(2), 74–93.

Meletiou-Mavrotheris, M., & Appiou-Nikiforou, M. (2015). Using models and modeling to support the development of college-level students' reasoning about statistical inference. In *Proceedings of the International Conference Turning Data into Knowledge: New Opportunities for Statistics Education, Lisbon, Portugal.*

Meletiou-Mavrotheris, M., Kleanthous, I., & Paparistodemou, E. (2014). Developing pre-service teachers' technological pedagogical content knowledge (TPACK) of sampling. In K. Makar, B. de Sousa, & R. Gould (Eds.), *Sustainability in Statistics Education. Proceedings of the Ninth International Conference on Teaching Statistics (ICOTS9), Flagstaff, Arizona, USA.* Voorburg: International Association of Statistics Education.

Meletiou-Mavrotheris, M., & Paparistodemou, E. (2015). Developing students' reasoning about samples and sampling in the context of informal inferences. *Educational Studies in Mathematics, 88*(3), 385–404.

Millar, A. M., & White, B. J. G. (2014). How do attitudes change from one stats course to the next? In K. Makar, B. de Sousa, & R. Gould (Eds.), *Sustainability in Statistics Education. Proceedings of the Ninth International Conference on Teaching Statistics (ICOTS9), Flagstaff, Arizona, USA.* Voorburg: International Association of Statistics Education.

Muis, K. R., Franco, G. M., & Gierus, B. (2011). Examining epistemic beliefs across conceptual and procedural knowledge in statistics. *ZDM Mathematics Education, 43*(4), 507–519.

Nacarato, A., & Grando, R. (2014). Teachers' professional development in a stochastics investigation community. In K. Makar, B. de Sousa, & R. Gould (Eds.), *Sustainability in Statistics Education. Proceedings of the Ninth International Conference on Teaching Statistics (ICOTS9), Flagstaff, Arizona, USA.* Voorburg: International Association of Statistics Education.

Nasser, F. M. (2004). Structural model of the effects of cognitive and affective factors on the achievement of Arabic-speaking pre-service teachers in introductory statistics. *Journal of Statistics Education, 12*(1).

Neumann, D. L., Hood, M., & Neumann, M. M. (2013). Using real-life data when teaching statistics: Students' perception of this strategy in an introductory statistics course. *Statistics Education Research Journal, 12*(2), 59–70.

Nikiforidou, Z., & Pange, J. (2011). Risk taking and probabilistic thinking in preschoolers. In M. Pytlak, T. Rowland, & E. Swoboda (Eds.), *Proceedings of the Seventh Congress of the European Society for Research in Mathematics Education (CERME7).* Poland: Rzeszow.

Nolan, M. M., Beran, T., & Hecker, K. G. (2012). Surveys assessing students' attitudes toward statistics: A systematic review of validity and reliability. *Statistics Education Research Journal, 11*(2), 103–123.

Noll, J., & Hancock, S. (2015). Proper and paradigmatic metonymy as a lens for characterizing student conceptions of distributions and sampling. *Educational Studies in Mathematics, 88*(3), 361–383.

North, D., Gal, I., & Zewotir, T. (2014). Building capacity for developing statistical literacy in a developing country: Lessons learned from an intervention. *Statistics Education Research Journal, 13*(2), 15–27.

Oesterhaus, J., & Biehler, R. (2014). Designing and implementing an alternative teaching concept within a continuous professional development course for German secondary school teachers. In K. Makar, B. de Sousa, & R. Gould (Eds.), *Sustainability in Statistics Education. Proceedings of the Ninth International Conference on Teaching Statistics (ICOTS9), Flagstaff, Arizona, USA.* Voorburg: International Association of Statistics Education.

Olfos, R., Soledad, E., & Morales, S. (2014). Open lessons impact statistics teachers' beliefs. In K. Makar, B. de Sousa, & R. Gould (Eds.), *Sustainability in Statistics Education. Proceedings of the Ninth International Conference on Teaching Statistics (ICOTS9), Flagstaff, Arizona, USA.* Voorburg: International Association of Statistics Education.

Onwuegbuzie, A. J. (1998). Teachers' attitudes toward statistics. *Psychological Reports, 83*, 1008–1100.

Pajares, M. F. (1992). Teachers' beliefs and educational research: Cleaning up a messy construct. *Review of Educational Research, 62*(3), 307–332.

Parsian, A., & Rejali, A. (2011). An experience on training mathematics teachers for teaching statistics in Iran. In C. Batanero, G. Burrill, & C. Reading (Eds.), *New ICMI study series: Vol. 14. Teaching statistics in school mathematics-challenges for teaching and teacher education: A joint ICMI/IASE Study: The 18th ICMI study* (pp. 37–40). Dordrecht: Springer Science +Business Media B.V.

Pearson, T. (2014). Improving the perceived value and affect of statistics in elementary and middle school teachers through the development of pedagogical content knowledge. In K. Makar, B. de Sousa, & R. Gould (Eds.), *Sustainability in Statistics Education. Proceedings of the Ninth International Conference on Teaching Statistics (ICOTS9), Flagstaff, Arizona, USA.* Voorburg: International Association of Statistics Education.

Peters, S. A. (2011). A robust understanding of variation. *Statistics Education Research Journal, 10*(1), 52–88.

Pfannkuch, M., Arnold, P., & Wild, C. J. (2015). What I see is not quite the way it really is: Students' emergent reasoning about sampling variability. *Educational Studies in Mathematics, 88*(3), 343–360.

Philipp, R. A. (2007). Mathematics teachers' beliefs and affect. In F. K. Lester (Ed.), *Second handbook of research on mathematics teaching and learning. A project of the National Council of Teachers of Mathematics* (pp. 257–315). Charlotte, NC: Information Age.

Ponte, J. P. (2011). Preparing teachers to meet the challenges of statistics education. In C. Batanero, G. Burrill, & C. Reading (Eds.), *New ICMI study series: Vol. 14. Teaching statistics in school mathematics-challenges for teaching and teacher education: A joint ICMI/IASE study: The 18th ICMI study* (pp. 299–310). Dordrecht: Springer Science+Business Media B.V.

Posner, M. A. (2011). The impact of a proficiency-based assessment and reassessment of learning outcomes system on student achievement and attitudes. *Statistics Education Research Journal, 10*(1), 3–14.

Posner, M. A. (2014). A fallacy in student attitude research: The impact of the first class. In K. Makar, B. de Sousa, & R. Gould (Eds.), *Sustainability in Statistics Education. Proceedings of the Ninth International Conference on Teaching Statistics (ICOTS9), Flagstaff, Arizona, USA.* Voorburg: International Association of Statistics Education.

Pratt, D., Levinson, R., Kent, P., Yogui, C., & Kapadia, R. (2012). A pedagogic appraisal of the Priority Heuristic. *ZDM Mathematics Education, 44*(7), 927–940.

Ramirez, C., & Bond, M. (2014). Comparing attitudes toward statistics among students enrolled in project-based and hybrid statistics courses. In K. Makar, B. de Sousa, & R. Gould (Eds.), *Sustainability in Statistics Education. Proceedings of the Ninth International Conference on Teaching Statistics (ICOTS9), Flagstaff, Arizona, USA.* Voorburg: International Association of Statistics Education.

Ramirez, C., Schau, C., & Emmioglu, E. (2012). The importance of attitudes in statistics education. *Statistics Education Research Journal, 11*(2), 57–71.

Reaburn, R. (2014). Introductory statistics course tertiary students' understanding of p-values. *Statistics Education Research Journal, 13*(1), 53–65.

Rheinberg, F., Vollmeyer, R., & Burns, B. D. (2001). FAM: Ein Fragebogen zur Erfassung aktuller Motivation in Lern- und Leistungssituationen [A questionnaire on current motivation in learning and performance situations]. *Diagnostica, 47*(2), 57–66.

Roberts, D. M., & Bilderback, E. W. (1980). Reliability and validity of a statistics attitude survey. *Educational and Psychological Measurement, 40*(1), 235–238.

Saldanha, L. (2015). Using TinkerPlots software to learn about sampling variability and distributions as a basis for making informal statistical inferences. In H. Oliveira, A. Henriques, A. P. Canavarro, C. Monteiro, C. Carvalho, J. P. Ponte, & S. Colaço (Eds.), *Proceedings of the International Conference Turning Data into Knowledge: New Opportunities for Statistics Education* (pp. 64–73). Lisbon, Portugal: Instituto de Educação da Universidade de Lisboa.

Sánchez-Sánchez, E. A., & Gómez-Blancarte, A. L. (2015). La negociación de significado como proceso de aprendizaje: el caso de un programa de desarrollo profesional en la enseñanza de la estadística [Negotiation of meanings as a learning process: A professional development program to teach statistics]. *Revista Latinoamericana de Investigación en Matemática Educativa, 18*(3), 387–419.

Santos, R. (2013). Prospective elementary school teachers' interpretation of central tendency measures during a statistical investigation. In B. Ubuz, C. Haser, & M. A. Mariotti (Eds.), *Proceedings of the Eighth Congress of the European Society for Research in Mathematics Education (CERME8).* Ankara: Middle East Technical University.

Schau, C. (2003). Survey of attitudes toward statistics (SATS-36). Retrieved from http://evaluationandstatistics.com/

Schau, C., & Emmioglu, E. (2012). Do introductory courses in the United States improve students' attitudes? *Statistics Education Research Journal, 1*(2), 86–94.

Schau, C., Stevens, J., Dauphinee, T. L., & Vecchio, A. D. (1995). The development and validation of the survey of attitudes toward statistics. *Educational and Psychological Measurement, 55*(5), 868–875.

Schnell, S., & Prediger, S. (2012). From "everything changes" to "for high numbers, it changes just a bit". *ZDM Mathematics Education, 44*(7), 825–840.

Schön, D. A. (1992). The theory of inquiry: Dewey's legacy to education. *Curriculum Inquiry, 22*(2), 119–139.

Shaughnessy, J. M. (2007). Research on statistics learning and reasoning. In F. K. Lester (Ed.), *Second handbook of research on mathematics teaching and learning. A project of the National Council of Teachers of Mathematics* (pp. 957–1009). Charlotte, NC: Information Age.

Shulman, L. S. (1986). Those who understand: Knowledge growth in teaching. *Educational Researcher, 57*(1), 4–14.

Skott, J. (2009). Contextualizing the notion of 'belief enactment'. *Journal of Mathematics Teacher Education, 12*(1), 27–46.

Souza, A. C., Lopes, C. E., & Oliveira, D. D. (2014). Stochastic education in childhood: Examining the learning of teachers and students. *Statistics Education Research Journal, 13*(2), 58–71.

Staub, F. C., & Stern, E. (2002). The nature of teachers' pedagogical content beliefs matters for students' achievement gains: Quasi-experimental evidence from elementary mathematics. *Journal of Educational Psychology, 94*(2), 344–355.

Sturm, A. (2016). *Überzeugungen zur Anwendbarkeit von Statistik [Beliefs on the applicability of statistics].* Wiesbaden: Springer.

Swanson, T., VanderStoep, J., & Tintle, N. (2014). Student attitudes toward statistics from a randomization-based curriculum. In K. Makar, B. de Sousa, & R. Gould (Eds.), *Sustainability in Statistics Education. Proceedings of the Ninth International Conference on Teaching Statistics (ICOTS9), Flagstaff, Arizona, USA.* Voorburg: International Association of Statistics Education.

Tintle, N., Topliff, K., VanderStoep, J., Holmes, V. L., & Swanson, T. (2012). Retention of statistical concepts in a preliminary randomization-based introductory statistics curriculum. *Statistics Education Research Journal, 11*(1), 21–40.

Vanhoof, S., Castro-Sotos, A. E., Onghena, P., Verschafel, L., van Dooren, W., & van den Noortgate, W. (2006). Attitudes toward statistics and their relationship with short- and long-term exam. *Journal of Statistics Education*, *14*(3).

Varian, H. (2009 January). *Hal Varian on how the Web challenges managers*. Taken from http://www.mckinsey.com/industries/high-tech/our-insights/hal-varian-on-how-the-web-challenges-managers

von Glasersfeld, E. (1993). Learning and adaptation in the theory of constructivism. *Communication and Cognition, 26*(3/4), 393–402.

Watson, J. M. (2001). Profiling teachers' competence and confidence to teach particular mathematics topics: The case of chance and data. *Journal of Mathematics Teacher Education, 4*(4), 305–337.

Watt, H. M. G., & Richardson, P. W. (2015). A motivational analysis of teachers' beliefs. In H. Fives (Ed.), *Educational psychology handbook series. International handbook of research on teachers' beliefs* (pp. 191–211). New York, NY: Routledge.

Wessels, H. (2014). Developing statistical knowledge for teaching of variability through professional development. In K. Makar, B. de Sousa, & R. Gould (Eds.), *Sustainability in Statistics Education. Proceedings of the Ninth International Conference on Teaching Statistics (ICOTS9), Flagstaff, Arizona, USA*. Voorburg: International Association of Statistics Education.

Wild, C. J., & Pfannkuch, M. (1999). Statistical thinking in empirical enquiry. *International Statistical Review, 67*(3), 223–248.

Williams, A. S. (2010). Statistics anxiety and instructor immediacy. *Journal of Statistics Education, 18*(2), 1–18.

Williams, A. S. (2013). Worry, intolerance of uncertainty, and statistics anxiety. *Statistics Education Research Journal, 12*(1), 48–59.

Wise, S. L. (1985). The development and validation of a scale measuring attitudes toward statistics. *Educational and Psychological Measurement, 45*(2), 401–405.

Wroughton, J. R., McGowan, H. M., Weiss, L. V., & Cope, T. M. (2013). Exploring the role of context in students' understanding of sampling. *Statistics Education Research Journal, 12*(2), 32–58.

Xu, W., Zhang, Y., Su, C., Cui, Z., & Qi, X. (2014). Roles of technology in student learning of university level biostatistics. *Statistics Education Research Journal, 13*(1), 66–76.

Yoclu, A., & Haser, C. (2013). 8th grade students' statistical literacy of average and variation concepts. In B. Ubuz, C. Haser, & M. A. Mariotti (Eds.), *Proceedings of the Eighth Congress of the European Society for Research in Mathematics Education (CERME8)*. Ankara: Middle East Technical University.

Yoon, K. S., Duncan, T., Lee, S. W.-Y., Scarloss, B., & Shapley, K. (2007). Reviewing the evidence on how teacher professional development affects student achievement. *Issues & Answers Report, 33*, 1–62.

Zapata-Cardona, L. (2014). A teacher development program in statistics within a community of practice. In K. Makar, B. de Sousa, & R. Gould (Eds.), *Sustainability in Statistics Education. Proceedings of the Ninth International Conference on Teaching Statistics (ICOTS9), Flagstaff, Arizona, USA*. Voorburg: International Association of Statistics Education

Zieffler, A., Park, J., Garfield, J., delMas, R., & Bjornsdotir, A. (2012). The statistics teaching inventory: A survey on statistics teachers' classroom practices and beliefs. *Journal of Statistics Education, 20*(1).

Zientek, L. R., Carter, T. A., Taylor, J. M., & Capraro, R. M. (2010). Preparing prospective teachers: An examination of attitudes toward statistics. *Journal of Mathematical Sciences & Mathematics Education, 5*(1), 25–38.

www.ingramcontent.com/pod-product-compliance
Lightning Source LLC
LaVergne TN
LVHW082240200926
848484LV00008B/2383

* 9 7 8 1 0 1 3 2 6 7 4 3 7 *